Surviving Common Core Math

West Chicago Public Library District
118 West Washington
West Chicago, IL 60185-2803
Phone # (630) 231-1552
Fax # (630) 231-1709

Surviving Common Core Math:
Finding Common Sense in the Confusion
2016 by Carol Pirog

All rights reserved. No portion of this book may be reproduced, stored in a retrieval system, or transmitted in any form or by any means (electronic, mechanical, photocopying, recording, or scanning) except for brief passages in critical reviews or articles, without the prior written permission of the publisher.

Published in the United States
by Anchor Book Press, Palatine
anchorbookpress.com

Text Copyright © 2017 Carol Pirog
All Rights Reserved
ISBN: 9781520357195

Surviving Common Core Math:

Finding Common Sense in the Confusion

By Carol Pirog

Anchor Book Press · Palatine

This book is dedicated to my mother, my first teacher, who taught me to love reading and math long before I started school. More importantly, as I grew older, she shared with me a love of writing that has lasted a lifetime for both of us. The publication of this book represents her success as much as mine.

Preface

Welcome to the world of education where things are not always done with the end user in mind. Many times, everyone jumps on the new idea bandwagon when there is no research to support the new ideas. While progress is made with new ideas, not every new idea is good. Sometimes it is better to wait and see if there is any reason to believe something works before everyone switches to the latest fad in education. I am continually amazed at the speed in which educators dump research-proven strategies while ignoring data, to follow the latest craze put forth by people who do not actually work with children. Educators are essentially throwing out the baby with the bath water. While this is a worn cliché, the mental picture is exactly what comes to mind.

This is the perfect analogy for the school setting, because we all know the dirty bath water needs to go, just as many things that are not working in education need to go. However, educated people should know to keep the baby – which I am translating to strategies that work while making changes that help students become successful, educated adults.

Further dirtying the water, political correctness causes administration to be quick to look at successful schools and then try to imitate what is happening in those school. That becomes a problem because it is not politically correct to look at the make-up of the student body to determine if the success is based on factors outside of the control of the local school, such as income, parental involvement levels and school readiness of students. It is not politically correct to say that a program or curriculum works for a group of students because their home life made them ready for school and academic tasks while the program would not be as successful with another group of students because the home life of those students has resulted in different learner characteristics for those students, which often requires different teaching strategies.

And most important, the people making the decisions are not in the classroom. They sit in their ivory tower and make decisions affecting your child on the faulty premise that what works for one child will work for every child. Having spent many years in the classroom, <u>I decided to write this book for the parents. However, there are many ideas that can be helpful for educators.</u> I have spent many years in the classroom, working with students from preschool through middle school years; students that were gifted, average, struggling or identified as needing special education; students from all economic backgrounds. The one constant I found across the spectrum – **one size never fits all, sometimes it doesn't even fit most.** <u>This book is to help parents who know that the elementary years are essential to their child's future, but are confused with the new Common Core Math.</u>

The good news for you is that **you are the most important variable in your child's education.** While schools can help all children, schools cannot replace involved and educated parents. **Good school or poor school, you can help your child succeed through the confusion!**

Surviving Common Core Math

How to Use This Book

There are main 4 sections to this book:
- Problems
- Goals
- Solutions
- Resources

You can start at any section, though I would recommend reading the solutions section before hitting the resources. Below is a brief highlight of each section to help you determine where to start reading.

Problems: This includes background on the Common Core and why it is a problem. It is interesting reading. If you are the type of person that wants to verify the why, this is definitely the place for you to start. However, if you and your child are struggling every night you might want to read this section after you have gotten started applying information in the solutions and resources sections.

Goals: Definitely important. While you may want to start with solutions, it is very important to be sure you know what the goals are, even if you wait on the problems. Often, when our children are struggling, we lose sight of the end goal while dealing with the present crisis. Your child's future is too important not to have the end goal in mind while you are working on solutions. Without the end goal focus, many short-term solutions effectively work against the end goal. This is a common problem in elementary schools because high school and college are so far away. Choices made during elementary school often close doors to higher educational opportunities because the end goal was not considered.

Solutions: Definitely the reason to buy this book. No problem starting here, but this is a short book and there is valuable information in the previous chapters. If you do start here, be sure to go back and at least skim the previous chapters. If you are the type of person who highlights things, this would be the chapter to highlight. As you will hear multiple times in this book, one size doesn't fit all, write in the margins; note what works or doesn't work for your child.

Resources: This section includes valuable resources for your child and more information for you. Again, one size doesn't fit all, so choose what works for your child. Try something; if it is helpful continue until it is no longer needed. If it doesn't work, try something different. I love the comment that defines insanity as doing the same thing and expecting different results. If it doesn't work, don't continue a strategy and expect different results. Finally, this is not an exhaustive list of resources, just something to get you started. As you become more comfortable you can look for additional resources. Search the internet, ask a teacher, or do some personal research asking other parents which resources were helpful for their children.

Table of Contents

Problems with Common Core 15

Goals to Survive the Common Core Package 33

Solutions 41

Synopsis 67

Teaching Math Facts 71

Internet Sites 79

Vocabulary Lists 85

Math Games 91

Problems with Common Core

There is a lot of hype out there about Common Core Standards, especially Common Core Math. Some of it is just rumors, a lot of it is true; often it is hilarious or would be if your child's education wasn't at stake.

The idea of Common Core was good, but like most things involving government bureaucracy something happened along the way. The idea was that educators would come up with a set of 'common' standards that all states would teach. The theory being that students in California would receive the same education as students in Illinois, as students in Alaska, as students in Kentucky, etc. The idea was good, so it was easy to sell it to educators and the public.

The first problem with Common Core is it became a package deal; rather than do one thing at a time, education reform was thrown in as well. As many people already know, **education reform is tied to high stakes testing. Then there was the new curriculum and text-books** that are required any time changes are made in when or what is taught. All of this created the **Common Core Package**, which includes the actual Common Core Standards, education reform, high stakes testing, new curriculum, and new textbooks. Then the package was presented to schools as the way to solve all their achievement problems.

Anytime anything changes in education, new text books are needed with the latest information. So, curriculum changes are thrown in the mix. Without changing the curriculum, it would be hard to convince cash-strapped schools to spend more money on new books. But not only are publishers selling a book, they are selling a program with services. Long gone are the days when schools just ordered student books and a teacher's book. Now there are whole packages with manipulatives, videos, computer components, assessment guides, and professional development. Professional development is required because the curriculum is so involved it is difficult for even experienced teachers to figure it out without costly professional development.

Then we have the educational reform and the resulting high stakes testing. Somewhere, somehow, in this process it was decided if high stakes tests were made harder; that would make the kids smarter. So, reform focused on high stakes testing. I don't know how educated people could come to the conclusion that the solution to low achievement and students unable to pass standardized tests is to make harder tests. Seems as if the people leading education should know that you don't address the symptom, you address the cause. So, reform and testing are included in the many things referred to as Common Core. In fact, many of the complaints about Common Core are really complaints about the testing.

That said, don't be too concerned about your child's scores on Common Core Achievement Tests; commonly called PARCC tests. I could write a whole book about problems with PARCC tests, but I'll just mention a few important concepts. It is rumored that many **good** teachers were having difficulty with the tests, so don't be dismayed if your child is also having difficulty. As we will cover later, what your child learns is important, not doing well on a test that may not be accurately assessing your child's knowledge on a particular topic. It is important that your child understand math skills and concepts. Just because they can't prove that understanding on specific tests does not mean they do not understand.

While new standards require new testing, it seems like the best idea would have been to figure out why the students couldn't pass the easier test rather than just making the test harder. Common sense would indicate that the cause of low scores would need to be addressed before we expect the students to pass a harder test. I keep asking myself is this the best we have to offer in the way educational leadership.

Interesting enough, if you look at the Common Core Standards by themselves, they really aren't that different than the old standards. The biggest change was the grade level when skills and concepts are taught. For example, Common Math Principles are being touted as an important addition to the way we teach math. If you look at the Principles, they include things like students will make sense of problems and persevere in solving problems and students will attend to precision. I don't know about everyone else, but I never had a math teacher that said it was ok to give up or that it was ok if my answer was almost right. As a teacher I certainly emphasized these principles way before Common Core was ever thought of. Developing the Package was a long process with the best minds and in the end, we were left with **Common Core Package** which includes the standards, new ways of teaching, and new testing. Many adults have found unlimited humor and many parents and students have found unlimited frustration with the **Common Core Package**.

To develop the Common Core Math Standards a committee was formed that included some of the greatest math minds in education. The math committee set about to determine what math skills and concepts should be taught and in which grades they should be taught and even how they should be taught in some cases.

This is the second problem with the Common Core Math Package – a look at the curriculum shows the textbooks were written by some of the greatest minds in math. While the majority of the authors should have been those with great math minds, they neglected to include those who struggled with math in school. A better idea would have been to choose at least some people who are successful in spite of early difficulties with math. This is critical to solving the problems with math education today. As a result of going with only 'math people', the people creating the textbooks and deciding when concepts should be taught and how long it should take to master those concepts are the people who breeze through math. The people who never experienced the frustration of just not understanding the concept the teacher is explaining. The people who did not need a lot of review and practice to master or remember a skill. The people who love math, who do math for fun, who enjoy solving difficult math problems.

Now the best math minds would have been a great idea, if the idea was to increase achievement of those students who are naturally gifted in math. However, if we stop to think about it, those mathematically gifted students were not the reason educators were calling for common standards or educational reform or improved test scores. No, the idea was to address the needs of students who were struggling under the current system, who were not making adequate progress under the old system, those who were unable to achieve an acceptable score on current tests.

With Common Core all is going well, until it is not. Think of it this way. If your child is like most children, they start out doing well in math. As preschoolers, they had no difficulty with math readiness skills. Things like understanding math concepts of less than and more than. Given the choice between 1 or 2 cookies, they knew that 2 cookies is better because everyone wants more cookies. They understood they wanted more French fries and less veggies. They understood inside and outside, big and little, first and last, same and different. They understood categories, knowing they had to pick 1 shirt and 1 pair of pants, not 2 pairs of pants when getting dressed. However, once they got to school they missed something. Who knows, maybe they were sick, maybe they weren't paying attention, maybe the kid in the next seat was being distracting, maybe they just had a great summer and didn't think about math until school started again. Bottom line, it doesn't matter. Your child missed or forgot important math skills or concept information. Now he or she is struggling in math.

Given the innate, linear nature of math, where skills and concepts build on previous knowledge, your child is now struggling because he or she does not understand the foundational skills and concepts needed to progress in math. In addition, given the structure of schools today we have multiple-choice tests and work in groups; no one is even aware that your child doesn't know the information for a few years; by that time, it is difficult for a teacher to figure out what your child doesn't know. When I taught remedial classes of 5 to 10 students, it was time consuming to determine the gaps in knowledge for each student. The task is almost impossible for the teacher with 20 or 30 students. To make matters worse, the new Common Core pacing guides found in textbooks, were created by math gurus who never needed to review and who have no idea how long it takes for the average math student to master a skill or concept. In the past, all math books had a chapter or two that reviewed previous math concepts. This was great for the students who forgot concepts over the summer or never mastered a topic. New textbooks make no allowance for time to review concepts and skills that were forgotten or never learned.

In the new curriculum/textbooks, multiple choice tests are another issue. Multiple choice problems allow students to miss foundational knowledge because many bright students are able to answer some questions correctly when they have choices. Actually, one testing strategy taught to students is 'guess and check'. Multiple choice tests allow students who have no idea how to solve a problem to get the answer right. As a result, even teachers who analyze student errors are not aware that the student does not understand the concept or skill. In addition, multiple choice answers often guide a student by eliminating some wrong answers. As a side point, I taught special education for years. One of the common strategies that is suggested to help students receiving special education services is to use multiple choice. I found that in math all it did was muddy the waters. I never knew whether a student knew something or not. If he got the problem right, did he know it or was it just a lucky guess. If he got it wrong, did he not know or did he just choose an answer without trying to solve the problem because it was almost time for recess?

This circles back to the problem of Common Core Math textbooks created by math gurus, the possibility of students not knowing prerequisite information is not accounted for. In the old days, the beginning of every year was reserved for review because in the old days we knew kids forgot things in the summer or they missed things in previous grades. If students were struggling during the review at the beginning of the year, the teacher knew that the students had not mastered the prerequisite concepts. In current pacing guides, even if the teacher is aware of skills and concepts that students are still struggling with, there is no time allowed to go back and re-teach prerequisite knowledge.

Now I hear things like, 'students should be able to add unlike fractions in the beginning of 5th grade because the last unit at the end of 4th grade was on equivalent fractions'. Me, being the skeptical person I am, think things like, 'you really think all the kids remember what we taught last May when all they could think about was getting out to play in the nice weather', or 'you really think the teacher got to the last unit in the book or had enough time to be sure concepts were mastered' or 'you really think that every child mastered all the concepts'.

Again, common sense says it doesn't matter that students should be able to do it. If they are not able to do it, they will struggle to process new information without knowing prerequisite information. Then, the students will get even farther behind because adding fractions with unlike denominators will never be completely understood as students struggle to solve problems, going step-by-step and never really understanding or reaching the mastery stage.

Group work that new textbooks favors, also contributes to the problem. While there are a lot of good things to say about students learning to work on a team, there are problems that need to be addressed. For example, many teachers see a group product and assume the entire group of students knows the information. Not only does group work prevent a teacher from seeing errors, it also prevents her from seeing when the problem is a lack of prerequisite knowledge. I have even seen teachers give tests in a group setting where students help each other (really students are just telling each other the answers) and then the teacher is surprised when told that testing must be individual work, not group work.

This brings us to the third problem – not only were the textbooks created by gurus, the actual Common Core Math Standards were developed by math gurus. Again, a group of people who never struggled with math. Basically, the committee did a good job of creating the standards. If you put the old and new math standards side-by-side, there isn't a lot of difference between the old standards and the Common Core Math Standards. The biggest problem comes with pacing and lack of review.

While they were creating the standards, an important idea was that **students should understand what they are doing in math rather than math being totally about engaging in rote memorization**. While many of us still remember that we did a lot of things, like invert and multiply, because we were told to, we did not understand why we were doing that. Correctly, the committee developing the Common Core Math Standards set about to correct this, saying kids needed to understand the concept and not just memorize the formula or rule. And that was a good thing. However, **what really happened was to make more math for students to learn as students must learn 'new' concepts in order to show they understand.** As math gurus tried to emphasize understanding, they actually confused students, parents, and teachers. This is the reason there are so many jokes out there about Common Core Math. If the strategy to teach understanding is more confusing than the concept, educators haven't really accomplished anything but to make more ideas that are confusing for students who were already struggling in math. Common sense would say to simplify if students are struggling, but the Common Core Standards are creating more confusion.

Common sense says that if you are having a child do something to help him or her understand a math skill or concept, it should be intuitive. If it is not, all you have done is create an activity that confuses the child even more. Take the idea of a drawing or visual representation to help students understand a concept. This was done well for years. Look at any old math book. If the concept is subtraction, there will be pictures of objects with some of the objects crossed out. If the concept was multiplication, there will be pictures of equal sized groups. I was teaching 5th grade math when Common Core was introduced. To show understanding of the concept of multiplying and dividing fractions, students were expected to create drawings. This created all kinds of problems which are based on students proving they understand, not the actual Common Core skill. Under the old standards, students were expected to multiply and divide fractions. Teachers were asked to take a test that would be given to students at my school. The test indicated that the majority of the teachers were confused as to which diagram represented multiplication. With the emphasis on understanding in Common Core, the text book and assessments are asking students which represented multiplication. If diagrams are supposed to help, it should be clear which diagram was needed. So, not only does your child need to learn how to multiply and divide fractions, they need to figure out which diagram to use creating more confusion.

The funny thing was, it wasn't even that hard of a concept, if only the text has used small fractions for the diagram. If they used fractions like ½ and ¼, at least the teachers would have been able to figure it out based on what makes sense. Although, they are using what they know about multiplying fractions to figure out the drawing. The drawing is not helping them to understand. Then you add the kids who can't draw a straight line or have visual perception issues and can't divide a shape into equal parts and we are bordering on mass hysteria. Add to that the fact that many times they are asking a kid to draw a shape with 5 or 7 or even 23 equal parts and they are just asking for trouble and frustration.

Think about this example, Common Core Math Standards say students will multiply using place value, diagrams, arrays, and the standard algorithm. The idea behind this is that it will help students understand and it allows students to use different methods to find the answers. The problem with this is students are required to prove that they know how to use all strategies for multiplying numbers. It has been my experience that the students who struggle to understand the concept of multiplication confuse and mix different strategies. It would be best to teach these struggling students one strategy. If a student is not able to understand with that strategy, then use a different strategy, not insist that in addition to learning a way to multiply the student must now learn all the ways to multiply.

Not only does teaching all the strategies take a lot of time, who needs to draw an array or diagram for a larger number? Once you understand the concept of multiplication, there is no need to draw a diagram for larger numbers. If you understand what multiplying is for 3 x 4, then you will understand what it is for 25 x 12. If you don't understand, drawing arrays with bigger numbers will only add to the confusion and result in less time available for learning the concept. All the time that it takes for teachers to teach multiple strategies and for students to draw diagrams and arrays is time that students could be using more productively to understand other concepts they are struggling with. Sure, use these strategies if they help a student to understand, but don't require students to learn them if they are just more confused as a result. In addition, don't waste a bright student's time, who doesn't need them in the first place as this takes valuable time this student could be learning more difficult concepts.

I was going to say finally, but this really isn't the last problem. It is just the last I am going to talk about in this chapter. So, **the next problem is the idea of one-size fits all in the Common Core Math Package.** This is really about interpretation of the standards and the curriculum. Just as we all know that one-size fits all t-shirt doesn't fit many of us, so the one-size fits all Common Core Math Curriculum does not fit all our kids. The one-size fits all comes into play with the curriculum, not the actual standards. Again, the biggest problem with Common Core is the **Common Core Package**, not just the standards. What we teach one child and how we teach one child is expected to work for every child. Not that the curriculum doesn't say it is differentiating for students with different needs, the curriculum does profess to do that. However, it is more like a Band-Aid and does not really address the needs of struggling students.

Along with the one-size fits all Common Core Math, the idea of **discovery learning or problem based learning that is being pushed in the one-size fits all curriculum.** Again, this is a problem with the Common Core Package, not a problem with Common Core Standards, but it is being addressed here because many people, including educators, do not see the difference between the actual standards and the curriculum or text book chosen by their district to teach the Common Core Standards. The discovery method/problem-based learning is a problem because of the one-size fits all curriculum, every child is expected to learn this way. However, this type curriculum has a number of issues.

First, struggling students do not learn well with the discovery method. Research has shown struggling student are most successful with explicit instruction in concepts and skills before they are asked to solve problems using those concepts and skills. For some students, all of math is so confusing, they need explicit instruction in all math topics. For other students, they are struggling with only one topic or concept and would benefit from explicit instruction in this one area while continuing in problem-based instruction for other math topics. I have taught struggling students, using explicit instruction, the same skills and concepts from the Common Core as were being taught with the general curriculum. It is possible to teach the Common Core Standards to struggling students as long as you use strategies that are proven successful with struggling students. Educators often confuse the method of teaching and the goal of teaching. The goal of students understanding math is so they are able to use math to solve problems. This does not mean the only way to teach math is with problem-based learning. Teachers are also confused because of all the hype out there about making the curriculum relevant to students. They think by putting the problem in a real-world situation they are making the problems relevant to students. The reality is the problems are often such a stretch to connect to students' life that it only confuses the students and does nothing to make math relevant to students.

Second, problem-based learning takes more time than it takes to explicitly teach students a skill or concept. Likewise, discovery models take more time than explicit instruction. While the discovery method is good for students who are not struggling. Students who are struggling cannot afford the time. They are already behind, they need to learn skills and concepts quickly as possible till they catch up. I don't know when we turned the corner in education, but now it is considered inferior to actually teach something explicitly rather than have a child discover the information on his or her own. This doesn't have to be a long-term solution. Maybe a child doesn't usually struggle, but is having difficulty with a particular topic. Generally, in a discovery model or problem-based curriculum, students work in groups. In most groups, the higher achieving students just tell the answer to a student who doesn't understand. So, the teacher thinks the student understands and when the student performs poorly on the test, the assumption is the student tests poorly, not that the student scored poorly because there was no one to tell him or her the answers.

Also, teachers fail to keep in mind that some students need more time to review a skill or have prerequisites retaught, but that time is not available if it has been used for problem based learning or discovery learning. Before you think I am against problem based learning, I am not. I just believe that it should not be used to teach **all** basic skills to **all** students **all** the time. The inflexible one-size fits all approach has caused it to become an issue by pushing discovery model and problem-based learning strategies for every topic and every student. Again, this is the problem with having only math gurus and not including people who are experts in how struggling children learn.

Third, some students can solve a specific problem, but are unable to generalize the skills for use in a different type of problem. Again, these students turn in classwork and homework that is correct because it is the same type of problem. When the test uses a different problem or different wording, the student is no longer able to answer the problem because he cannot take what he learned in one situation and use it in another. Along that same line, when the assessment is a problem based test, it is difficult to determine where the student is struggling. Is the issue with prerequisite knowledge needed for the problem, is it lack of understanding of the skills and concepts being tested, is it difficulty applying the skills and concepts to a real life problem or is it a reading problem and the student didn't understand the problem.

In summary, there are problems with Common Core Math Standards, but the reality is the **Common Core Math** Package which includes standards, education reform, testing, and new curriculum is a bigger problem. When looking for solutions for your struggling child, try to separate which of these issues is causing the problem for your child. Your child needs to know and be able to apply math skills and concepts, in other words the Standards. When your child gets to high school, he or she will not be asked 5 ways to divide or to draw a picture. In the end (high school and college) knowledge of standards is what is important, not PARCC scores, not knowing 5 ways to divide, not being able to draw a picture. As your child continues his or her education, focus on being sure your child learns the skills and concepts. Don't worry as much about the fluff. Knowing the skills and concepts will enable your child to survive Common Core educational swings through the popular theories of the years.

As a side note, I know of at least one state that has already gotten rid of the new testing at the high school level. In Illinois, it was announced last summer that the state would no longer give the PARCC test to students in their junior year. Instead, students would be given the SAT. So, the revision of the **Common Core Package** has begun. It will probably take years to reach the elementary level. By that time your child will be long gone. However, all is not lost if they arrive at high school with a firm knowledge of the skills and concepts needed to be successful.

Goals to Survive the Common Core Package

Before we think about how to solve any problem, we need to think about the goals we want to achieve. My first thought was to discuss the goals in order of importance. However, every time I tried to decide which was most important, upon reflecting on the consequences of not meeting the other goals, I had to change my mind. So, the goals are presented in a random order because they are all essential.

The first goal is education. I know this seems like a 'duh' goal, but as a teacher I have been in lots of meetings, discussions, and professional development where it seems as if it has been forgotten that student education or achievement is what school should be all about. This is not to say that there aren't a lot of other worthy goals. However, sometimes these goals are conflicting. Bottom line, parents have an unwritten expectation that their children are going to school to receive an education.

What does that mean with the Common Core Math Package? Education is not about being successful in any given curriculum or way of learning. It is not even being successful on any given type of test. <u>It is about learning the skills and concepts that your child will need to be a contributing member of society as an adult</u>. For math, it means your child will master math skills and concepts needed to manage a household and a family. Everyday things like measuring the ingredients for a recipe, cutting that recipe in half or doubling it for parties. Things like going to the store and determining which product is a better buy considering cost and quality. Paying bills and choosing goods and services based on the best economic value for the money. Having a checking account and avoiding overdraft charges. Knowing the importance of long-term savings and how to invest those funds. Now, I know the first thought of many is to say that limited skills are needed as calculators are readily available, but using a calculator is not a substitute for learning elementary math calculations.

The calculator debate is still going strong in education circles. Here are some facts about calculator usage that even many teachers are unaware of. First, the students who are the worst at math often score no better when they are allowed to use a calculator than they do when they do not use a calculator. The reason is that everyday math is not based on only the ability to correctly calculate addition, subtraction, multiplication and division problems. To solve every day problems, it is important to know which operation or operations you are going to use, and in which order you will use them. For struggling students, those 4 choices baffle them. They do not know which to choose. If you ask any teacher, they will tell you that the biggest area of difficulty for most students is word problems, basically figuring out which operation to use.

This goes back to the number sense and basic math skills that they began learning before they started school and in kindergarten and first grade. It is also part of the problem that Common Core is trying to solve, even though they are largely unsuccessful due to the methods they are using. Students do not understand what they are doing when they add, subtract, multiply or divide. Common Core attempts to address this problem by having students use diagrams and pictures. While this is a great idea, it only works if the drawings are intuitive. By this, I mean that you would know what is going on by just looking at a picture. However, once we get past the basics of counting and adding and subtracting whole numbers, many pictures and diagrams are way too confusing to help students understand the concepts.

The second calculator issue is for students who are a little better in math. In real life, they know they should add to find the total cost for their items, then to subtract to find out how much money they will have left after a purchase. What they are lacking is the number sense to know they have made a big error. Without a strong number sense and knowledge of basic math facts, when a wrong key is hit on the calculator, they are unaware that the answer cannot be right. After word problems one of the biggest difficulties students have in school is estimating the correct answer. This is a life skill that is missed when the calculator is used. I have seen many students who find the answer on a calculator, then round in order to give an answer to a problem asking the student to estimate. In this case, a calculator defeats the purpose of the student being able to estimate the correct answer.

The third issue with the calculator is overuse. Leaders in math education are pro-calculator use and parents, students, and teachers take this as a blanket endorsement of giving kids calculators. However, if we look into it more closely, leading math educators believe calculators should be used for more complex problems, not basic facts. While there is some disagreement over how quickly a student should be able to spit out math facts, none of the leaders in math education are suggesting that students should be given a calculator to figure 6 + 4 or 3 x 5. These are the facts needed to estimate your answer.

Therefore, a calculator will not meet the needs of your child to become a competent adult, able to use math in everyday life. Education should provide students with the knowledge and ability to apply math skills and concepts in daily life. As parents it is our job to see that happens.

The second goal is to allow your child to go to college or a technical school after high school. There is the possibility a child will go to a trade school, enroll in a certificate program with no math involved, or go directly into the workforce after high school, but we should not limit a child's choice when they are in elementary school unless the child has a mental impairment that would definitely prevent them from going to college. For all students with average intelligence, the goal is to make sure that elementary education prepares them for high school algebra and geometry which are needed to take college-level math courses.

I cannot even count the number of students I know that have had to take remedial math classes in college. For those of you who do not have students near college-age, these are courses offered by some colleges that do not count as college credit. They cost the same as college level courses. Students who are unable to pass the college math placement tests are required to pass the remedial math courses before they are allowed to enroll in math for college credit. The community college near me offers 4 levels of math covering pre-college math. It is great that these courses are offered, allowing students a second chance. However, in a best-case scenario they add to the time your student spends at college, often making a 2 or 4-year degree take 3 to 5 years to complete. And in this case, time is money. The longer a student is at college the more money it takes. The worst-case scenario, for some students it has made the difference between completing college and dropping out. Many college-age students do not have the stamina and drive required to learn all the math they should have known when they completed high school

The bottom line is that if your child needs a calculator to figure out basic math facts they are unlikely to have to stamina needed to complete a college level math class. Using a calculator for basic facts adds to the time it takes to solve a math problem. There is the possibility that your student will have teachers in high school that will hold their hands and get them through high school math, but in college it is a whole different ballgame. The level of support is not nearly as great, nor should it be. The goal of parents is to be sure educators help students prepare, knowing the foundational skills and concepts to allow them to be successful in high school and college. For most people reading this book, I am preaching to the choir. Higher education is an essential goal. Without it, many of your child's choices will be eliminated.

Another goal that we cannot forget is your child's self-esteem. Self-esteem has gotten some negative press in the last few years. Don't confuse this goal with the negative idea of a student feeling proud of minimal accomplishment. I'm talking about a child having a positive self-concept that allows him or her to face challenges without giving up because they know they are capable of learning. Remember, it doesn't help anyone for your child to think that he or she is just **not** capable of learning math skills and concepts for the grade they are in. Maybe they won't understand with the method they are being taught, but they can understand, and your child needs to know that they are able to learn a skill or concept. It is ok for them to learn in a different way, and they need to know that all people learn differently. There is nothing wrong with that. Wouldn't the world be a boring place if we were all the same? Education is no different. We don't all learn the same way, but some schools insist on teaching everyone the same way. Just as we wouldn't say everyone has to wear the same clothes or go to the same church or listen to the same music, so we should not insist everyone learn using the same teaching method. Whatever decisions that are made to solve your child's struggles with the Common Core need to be made in a way that protects your child's self-esteem. The key to developing or maintaining self-esteem is **success**. Nothing bolsters self-esteem like succeeding. Our goal as parents is to be sure that our child has the chance to succeed by being sure they are taught in a way they can learn which will go a long way to creating the self-esteem needed to engage in challenging tasks.

Finally, we cannot forget the goal of relationships. We do not want math education to be so difficult that it practically destroys our relationship with our children. Every night should not consist of a routine of tears and harsh words while helping our children with their homework. This is often the forgotten side of education, but it is very important, especially in today's world. There are so many other things to attract our child's attention, we do not want a large portion of the time spent with us to be confrontational. Yet, we cannot just ignore math in favor of creating a more pleasant climate at home. There needs to be a way to do both, help our children complete their homework successfully and to enjoy time with our child that fosters the parent-child relationship. If you and your child are already in a confrontational relationship, consider a tutor or check out some books on parenting styles. Books like *Parenting with Love and Logic* describe parenting styles that avoid confrontation without endorsing permissiveness.

In conclusion, there are 4 basic goals of surviving the Common Core Package. First, you children should receive an education that allows them to function as adults. Second, they need to learn enough math that they are able to complete high school and college math courses in order to pursue a career. Finally, this needs to be done in a way that does not destroy your child's self-esteem or your relationship with your child.

Solutions

There **Are** Ways to Help Your Child

Before we start on solutions, we need to review 3 essential facts that affect available solutions.

1. **Math builds on previous knowledge**. Any solution that does not recognize this crucial fact is doomed to failure. Always be aware of prerequisite skills.
2. **Schools teach by grade level**. It is important remember that no matter where your child is at in current math achievement, the school will be teaching at grade level. Many times, this is true even in remedial programs. If you are unable to get your school to address previous knowledge, you will need to do it yourself or hire someone else to help your child master prerequisite knowledge if your child is to become successful.
3. **Students must know basic facts**. This means that your child should know multiplication and division facts by memory AND understand what multiplication and division mean. Research has shown that students who make the most progress in math learn their basic facts as they learn math skills and concepts. One is not more important than the other; they are intertwined, promoting a solid foundation of understanding. If your child is in the 4th grade or higher and does not know multiplication and division facts, start teaching them now! Make it a game,

give rewards, do whatever is necessary. It is ok if your child still uses their fingers for addition and subtraction, but multiplication and division must be memorized. It is truly the basis of many higher math skills, as multiplication and division show the relationship between numbers. The use of a calculator for basic math facts makes it harder for the student to grasp the relationship of numbers. The Resource Section offers some suggestions for helping children learn basic math facts.

Now that we understand some of the problems of Common Core Math instruction and goals for our children, let's look at what can be done to help your child survive the Common Core Math Package. Just as reading instruction swings back and forth from whole language, to phonics, to balanced literacy, so Common Core Math instruction will swing back to include the basics. Actually, **a balanced math program includes rote memory work, skill instruction, and application/problem solving.** The important thing is to help your child learn with a balanced program until the Common Sense Swing occurs. Some children will be better at memory work and other children will excel in application of skills. But all children need to be able to do both.

I was discussing the Common Core Math issues with some high school math teachers and they really didn't understand what all the hype was about. Math hasn't really changed. Sure there are different ways to solve problems, but when your child gets to high school, they will basically be doing the same thing that has always been done in math. Sure, there will be more application of math and more word problems, but the math is the same. By high school, the teachers are no longer hung up on whether your child knows all the ways to solve a multiplication problem or if your child can draw a diagram that goes with a specific equation. Once your child gets to high school, they will be learning the same algebra and geometry that you learned. They will be asked to think mathematically. Thinking mathematically is the key, but your child will struggle to accomplish this if he or she does not have a solid foundation in math skills and concepts. As I am writing this, it has just been announced that Illinois will no longer give the PARCC (high stakes testing associated with Common Core) to juniors. Instead juniors will take the SAT at state expense.

There are a few things that you can do to make sure your child arrives at high school with the essential skills, his or her self-esteem, and your relationship intact. They will take time and/or money. But your child is worth it. The earlier you intervene the more successful the intervention will be and the less time and tears it will take. I say it will take time and/or money because you can do the work yourself if you have the time and the patience, you can hire someone or some organization to do the work for you, or you can use a combination. With my children I used a combination. Things like vocabulary and math facts are better taught for a few minutes a day, rather than once a week with a tutor.

To start with, you have to decide if you want to do the work yourself or seek help from an outside organization. One option is not better than the other. It will depend on your own skill level, patience, time, and budget.

As you start, one of the most important things for your child is to learn is **math vocabulary**. Don't assume that your child knows math vocabulary. I taught 5th grade students for 10 years and I was always surprised when I found many students did not know the meaning of the word **sum**. So, when a word problem said to find the sum, they often didn't know what to do. You can skim your child's math text to get specific words your child needs at their grade level. There are grade-level lists in the resource section of this book, but keep in mind the lists are general math vocabulary and the specific curriculum your child uses may include different terms. If you use the general lists, your child will be prepared for high school, but may continue to struggle now. You can just google math vocabulary lists. Some sites even have math flashcards that you can print. One caution, do not try to do all the words at once, this only adds to your child's confusion. Start with 2 or 3 words and add words as your child masters previous words. Be sure you are continually reviewing words, especially words that are not used every day.

If your child is struggling in math, you need to be sure they know the terms that were taught in previous years. I worked in low income schools for a number of years. I noticed that all struggling students had a lack of understanding of math terms. Many times, the student does not even realize they don't know the meaning of a word because they know other, non-math meanings of the word. A perfect example is mean. When your child uses the word, they are generally talking about someone's unkind behavior or comments. As you know, in math the word carries a very different meaning. Other times the child knows exactly what the word means and can give you a definition, but does not translate that into a math action or operation. All school age children that attend a regular school know what it means when something is taken away. However, I have seen countless students read a word problem where something in taken away and proceed to try to solve that problem by adding. They get the problem wrong, but it is a vocabulary error not a calculation error.

Another big reason to teach vocabulary is that students who do not understand math terms are quickly lost when the teacher starts explaining a new skill or concept. Most children do not raise their hands to get clarification, they just sit and listen or fidget or distract the student next to them. As a result, they don't know how to do the problem. Many times, the teacher will determine little Johnny doesn't understand because he doesn't pay attention when it is really the reverse. Little Johnny isn't paying attention because he doesn't understand what the teacher is talking about. As a parent, you attend parent-teacher conferences and it seems Johnny's lack of progress is due to lack of effort. However, it needs to be determined if lack of effort is the cause or the symptom. It can be either. Is your child confused because he isn't paying attention or is your child not paying attention because he is confused? In the end, the results are the same, but the solution is not.

I have even heard teachers tell a student to go figure it out. I am not going to repeat what I already told you. I have also heard teachers say they don't understand why the students are struggling because she told them how to do the problems. My experience has been in many cases it is tied to vocabulary. This is even truer if the child also struggles with reading comprehension. A student who receives speech and language services for receptive and expressive language, or a student who is a second language learner will most likely need help with math vocabulary. Everyone expects that, but it is true for many other students as well. If you haven't looked at a math book lately, you will be amazed at how much reading is involved.

One of the easiest ways to teach vocabulary is indirectly through math games. When you use games, and talk about what is going on, most children pick up on the vocabulary needed for the game. They need to know this information to win the game which provides motivation. Sitting in a class many students don't see a reason why they need to know the vocabulary or they have struggled for so long they don't feel they are capable of learning. If your child is into sports, talking about sports is a good way to get in basic vocabulary. Things like, look <u>how many more</u> points our team has. You know you subtract to figure out how many more. Or this was a high scoring game. Let's add the points from both sides to get the <u>total</u> points scored during this game. Many kids like to learn something they can show off to their peers. While we don't want our child to be annoyingly egotistical, it never hurts for a kid to know there is something he or she is the best at, especially if that student struggles in other academic areas. One word of caution, be sure to make it fun. If you feel yourself becoming annoyed because your child just doesn't get it, do something different for a while.

At times, you may need a more direct approach. The resource section has a list of websites to find vocabulary. As I already mentioned, do not introduce too many new words at once for a struggling student. A combination of games and direct instruction usually works best for most students.

There are many games that you can buy that involve math. I love the old standards like, Monopoly, Yahtzee and even Trouble which involves counting and strategies for younger children. There are games available specifically for math. The games listed above were created for families to have fun, improving math skills is a by-product. Search the internet for math games or go to a teachers' store. The resources section of this book includes some games that you can buy or make at home. The key for these games is to be sure you and your child talk so that the math vocabulary becomes part of their vocabulary. Remember, fun, fun, fun.

After vocabulary, the most important idea **is to teach your child the skills and concepts that are needed at their grade level**. There are a number of ways to do this. First, you can do it yourself. If you completed high school math with a reasonable level of understanding, you should be able to help your primary or intermediate student with skills and strategies even if you are totally baffled by the Common Core Math. It is not really the Common Core Math skills and concepts you are confused about. It is the discovery model curriculum that is confusing, or the different strategies meant to show that students understand the concept. As mentioned earlier, when your child gets to high school it will be important that they know specific skills and concepts, not a certain way of learning or solving problems. Focus on teaching the skill if you do not understand the strategy your child is learning at school. When you child says that isn't what my teacher said say, "I know, isn't it great there are so many different ways to learn math."

One of the biggest changes in the Common Core Math Standards from the old math standards was just the switching around of grade level expectations. In the old standards, students were expected to know basic multiplication and division facts in 4th grade. In the Common Core Standards, they are expected to know the multiplication and division facts in 3rd grade. Keep in mind, they are the same facts you learned in school. If your child didn't learn them in 3rd grade, teach them in 4th, or 5th or 6th. Except in a few instances, most of the problems with Common Core Math is with the interpretation of the standards and the curriculum, not the actual skills and concepts. If you stick with the skills and concepts, you and your child can survive and meet the essential goals of math education. Remember, one-size does not fit all.

You can help your child yourself with the help of some internet programs. There are many good programs out there. Some are free, and some have a subscription fee. There is a list in the resource section of this book of some of the sites available at the time this was written. A google search will lists many sites. Remember, try a site and if it is helping use it till your child no longer needs it. If it doesn't work, try a different site or a different approach altogether. One word of caution, the internet is not the answer for every student. Some students need a paper-pencil approach. In that case, use the internet sites that allow you to print the work and the student works offline with real paper and pencil or you can purchase other paper and pencil materials from websites or teacher stores. I have found that accountability is easier to maintain with paper and pencil tasks. If self-motivation is an issue for your child, you might consider starting with paper and pencil tasks. Another plus for paper and pencil is your child can see exactly how much they have to do. In addition, it is easier to be sure your student makes corrections. As a side point, one of the best strategies for learning is correcting mistakes. Correcting mistakes is a key as worksheets are worthless if the student does not learn from their mistakes.

Another option, you can arrange for private tutoring or enroll your student in one of the commercial programs available like Sylvan or Kumon. Private tutoring is often expensive and is only as good as the individual you hire. In addition, most tutors just cover what the child is learning now, leaving your child with the gaps in knowledge that will surface again. If your child has been successful until now, a good tutor might be just what your child needs. If your child has been struggling for multiple grades, a commercial program might be best.

Most commercial programs have a standard program with research to show results of the students who were in the program. They use a combination of your student being at their facility learning and your student practicing the skills at home. The only program I am personally familiar with is Kumon. Kumon has an excellent program. Your child needs to attend the center one or two days a week. Kumon will send home practice sheets for the 6 days he or she is not at Kumon. You will need to grade the sheets and have your child make corrections (there are answer keys available and this takes much less time than you would spend tutoring your child).

Also, this is less than the price of most individual tutors and they have a balanced plan, covering all math topics through high school. Instructors keep track of your child's progress and have them move along according to individual achievement. At Kumon, they will test your child and place your child so that your child learns the skills and concepts that he or she missed in school. While some educators criticize Kumon because it does not use a problem-based or discovery method, that is not their purpose. The purpose is to help your child master the skills and concepts that will allow them to be successful in a problem-based curriculum. Further, research has shown that explicit instruction is more effective with struggling students than discovery or problem-based learning. As a side point, Kumon and other tutoring centers are for profit organizations. They only stay in business if they make a profit. Unlike public schools, they must do what works in order to be profitable. Parents continue to pay the tuition because what their child is receiving for free in the public school system is not working for their child and the paid program is helping.

Unless you have a background in education or lots of time for research, finding the gaps in your child's math knowledge is one of the harder things to do. Even with a background in education it is time consuming to find the gaps. So, looking into an organization that is making a profit because your child is being successful is sometimes the best choice. However, if you do not want to start with that option or prefer to be involved even with this option, there are resources that can help.

I have given you some ideas how to get instruction for your child, but what to teach is just as important as where. Especially if you have decided that you will provide the instruction yourself. As a teacher, the first thing I always did when planning instruction was to look at the end of the unit test. What are the students expected to know when the unit is complete? Unfortunately, you will probably not have access to that information unless you are really good friends with the teacher. The next thing to do is look at the book. Ask for one if you child does not bring one home. If the teacher says there isn't one, ask for a list of standards that will be covered during the next unit.

While the standards are sometimes confusing, you can usually figure out the basic idea. If not, ask the teacher what they mean in terms of what type of problems your child needs to be able to do. You may need to insist on this information. I clearly remember a situation when I was an interventionist for students struggling in math. I was trying to get this information and the teacher did not want to give it to me. He was afraid that it would give the student an unfair advantage if I explained how to do the problems. Of course, I was clearly of the mind that the child should have every advantage possible in order to be successful. This was a former student. She had taken the initiative to come to me. She told me she was confused and did not understand when the teacher tried to help her. Turns out, I realized it was a vocabulary issue once I convinced the teacher to give me the information.

Third, **make sure your child learns concepts they may have missed in prior grades.** Unlike reading that teaches many of the same ideas with increasingly difficult text year after year, math continues to add new skills and concepts while expecting the student to know and use skills and concepts taught in previous grades. When a student misses some concept in a book the class is reading, it isn't as devastating as missing how to regroup or order of operations. In reading, the concept will be taught again, possibly in the next book, but definitely in later books and future grades. This isn't the case with math, once a student misses something and the class is finished with it, most teachers do not go back to reteach. However, the curriculum will build on that knowledge. This is one of the major problems with the curriculum for Common Core Math. As an interventionist for struggling students, I was always amazed at the number of students who could not read numbers in the 1000s. These were 5th grade students who were working with numbers in the millions and numbers that included decimals. How could anyone expect a child to comprehend math problems when they couldn't even read the number that was the answer? Yet, no one ever tested or retaught that knowledge. Not only that, I found all my students were able to master this foundational skill with just a few minutes of work each day.

To illustrate this in another example, in one 3rd grade Common Core Math textbook students spend a whole semester learning to multiply and divide and learning the multiplication and division facts. However, some teachers in 5th grade are shocked to learn that the Common Core Math standards includes memorization of multiplication facts because nowhere in the 5th grade standards or curriculum is it mentioned that students need to do this in 5th grade. It is assumed (by the math gurus) that students learned the multiplication facts in 3rd grade, no need to revisit them in later grades. At a professional development class taught by the publisher, a teacher asked what should be done about students who do not know the basic math facts. The response was there was no need to teach them again as students would just pick them up as they were working with the problems. My thought was that if they did not learn the facts when there were being taught, they were not going to learn them when something else was being taught. Again, another skill students can master with just a few minutes of practice each day. In my class, even students that many had already decided were incapable of learning basic facts just needed the right strategy to be successful.

It is my personal opinion that this is the problem of most people who say the just don't get math. At some point in their educational lives they missed a math concept or skill. While they don't really understand this concept, future math is being based on the concept. In schools, the curriculum is too jam packed for teachers to take the time to determine if students know prerequisite knowledge. And even if they did, there is no time allotted for re-teaching previous concepts. For example, the teacher is teaching the algorithm for long division. She will barely be allotted enough time to teach this skill to bright students who love math, using the discovery method. There is no way she can go back and assess whether each student can multiply and subtract. Or whether each student knows the place value system well enough to divide using the standard algorithm. Much less spend time teaching those students. So, as parents, it falls to us to make sure our students know the prerequisite skills for the current math.

Let's look at the classic problem of students not knowing their multiplication and division facts. Many teachers say to let the students use a calculator or multiplication chart, but this discounts the bigger picture. Much of math is about the relationship of numbers. I just know that 60 is twice as big as 30 or that 50 is ten times as big as 5. I know that because I know multiplication facts. I apply that knowledge when I am reducing fractions or when I am finding equivalent fractions or ratios. It is just way too cumbersome for a student to have to use a chart to figure out the multiples of 5 or 6.

Most people have heard of fluency in reading and it is considered a given fact that students who struggle with fluency in reading (reading with expression and quickly figuring out unknown words) have lower comprehension. The reason being that some of the thought processes that should be used comprehending are being used to figure out the words. The same is true for math, if a child does not know basic math facts, thought processes that should be engaged in solving the problem are being used for figuring out basic facts. If your child is in 3rd grade or higher, they will always struggle with math unless they learn their facts. There are some ideas for learning facts in the resource section of this book. You can also use internet sources or find books in the teachers' store.

One excellent resource if you are trying to fill in gaps is the **Drops in the Bucket** series. I love this series as it is not complicated with a lot of language. The concept is to teach the skills that are needed for problem solving and application. They are available through Frog Publications for less than $20 for each grade level. The books are blackline masters, which means you can make copies for your own use if you purchase the book. Each book is 60 lessons (which covers a year of math skills) and an answer key. With the new Common Core Standards, the grade levels do not always match up with current math topics, but your child still needs to know all the skills in the book. This is a true intervention in the sense that it is meant to help students catch up. There is not a lot of fluff. A child can cover the material for one grade level in about 3 months if they are doing one page a day.

Here are some tips if you decide to use this resource:
1. Use level for where you think your child is functioning, not current grade level.
2. If unsure of starting place, copy Lesson 10. If your child gets 90% or 100%, repeat with Lesson 20. Keep going in this manner till you find the right starting place. At whatever point your child doesn't get at least 90%, go back. For Lesson 10, start on Lesson 1. For Lesson 20 start on Lesson 11, etc. (Percentage is based on getting the topic right. For example, if number 5 has 4 problems, that counts as 1 wrong if all are not correct.)
3. Once starting level is determined, copy each day's lesson rather than have your child work out of the book.
4. Have your child complete the lesson. Check it immediately. Have child make corrections immediately. If child missed more than 2 problems, repeat same page the next day.
5. If your child continues to miss the same problem, find supplemental material for extra practice. (Common Core Sheets is a good place to start looking for printables or see the resource section for websites.)
6. If your child gets 100% on a **new** page 2 days in a row, skip 5 pages.

While this may not align with requirements grade-wise, it does teach math concepts that will give your child a firm foundation and increase their ability to think mathematically and solve application problems. This resource can also be used as a supplement to help a child exceed the skill level expected at their grade level. Again, this does not teach the problem solving required of Common Core Math, but teaches the skills needed for problem solving. It is organized around 10 math topics. Each day, problem 1 continues on the same topic. Likewise, number 7 will always be about the same topic. If your child is sensitive, but needs below grade level instruction, keep the book put away and cover any grade levels before you copy the worksheet. I had the most success when I provided some small reward for passing each 'level'. Usually I used Lessons 10, 20, 30… as quizzes with rewards. I used a bigger reward for completing a book as that is a year's worth of math skills.

The next thing is to **make sure your child is engaged.** No matter how good the program or how good the teacher, if your child is checked out they will not learn. There is a lot of debate about intrinsic and extrinsic motivation. Everyone knows that intrinsic motivation is the best. Your child will put forth way more effort for something they want to do than something they don't want to do. **Our job as parents is to provide the motivation a child is lacking.** There is a whole psychological debate in the area of motivation. So, let me just give you some pointers and some background. It may be common sense, but sometimes we forget the basics when our child is struggling.

First of all, be positive. It doesn't help to tell your child he or she is just not a math person and probably never will be, or even you aren't a math person. It is ok to commiserate with your child and say you remember the struggle, but you have to end on a positive note. You must have eventually figured it out or you wouldn't be helping them now. It is good if you can share how someone else helped you, this lets your child know that it is ok to get help from others. Sometimes kids are just embarrassed because they don't know or can't figure it out. It helps children to know others have the same difficulties.

If math was always easy for you, it is probably best not to point that out. No need to make the kid feel worse or more hopeless than he already does. Be encouraging, letting him or her know that you are confident they will figure it out. The worst thing you can do is make you child feel stupid. If you don't have the patience, it is better to spend the money to get someone else to help your child and try to make encouraging remarks as often as possible. If your child feels it is hopeless, it will add another barrier to overcome. In addition, it will make math an emotional issue. Research has shown emotional stress reduces the ability of the brain to learn and solve problems. There is some good research on this if you would like to know more about how emotions effect learning. There are links to several articles in the resource section

In addition to being positive, sometimes you need to provide external rewards. I remember when I first started teaching I was against any sort of external reward. After all, learning was its own reward, I thought. Why should I bribe students to do what they should be doing? However, it didn't take long for me to change my mind. At first it was just survival. In a low-income school where students and parents often did not value education, it was difficult to convince students to do their work because they should. I was going against the students' peers, family and current achievement levels. To work hard in school was looked down upon. The mindset of why waste your energy on something of no value was prevalent.

Sometimes the family did value education, but the students were already out of control. Or the student was struggling, and the parent was unable to help. The biggest factor was the difficulty. These were struggling students. The students in the gifted classes were barely at grade level and everyone else was 2 or 3 years behind. It was just not reasonable to think they would work really hard for a benefit that they did not see as valuable or attainable. They were already convinced that they would not be able to be successful and they had years of failure to back up their conclusions.

One day it clicked; I was going to work every day because I got paid, not because helping children was a nice thing to. While I did want to help children, if I was not paid at my current school I would have gone to another school where I would get paid. So, I went with the reward system. You have to create some success before you can reasonably expect intrinsic motivation to be there. The best rewards are not tangible, things like spending time playing a special game or extra time at the park or watching a TV show. Sometimes, the difficulty is so great there does need to be a tangible reward like a small piece of candy or small toy. It is best to refrain from punishment as that tends to highlight failure. Plus, there is always the chance you could be punishing your child for something he or she is just not capable of doing this week. In our family, even negative consequences were presented as a choice. You can play a video game when your work is done. Or you can watch the Disney channel as soon as we finish this math review sheet. We also explicitly taught that choices have consequences. Good choices lead to good things. Poor choices lead to negative things. For example, not being allowed to watch a movie was not presented as a punishment, it was just that you don't have time if you have not completed your assignments. You can find more about this concept in the book *Parenting with Love and Logic.*

The next concept is **accountability** – yours, your child's, your child's teacher. **For you**, be consistent. Commit to helping your child be successful. Buy math games and make them fun. If you need to buy a game meant for a younger child so that your child enjoys the game, do it. Everyone loves easy games, they build on the idea that math is fun, and they promote enough self-confidence for your child to work to solve more difficult math problems. Do math things around the house, for younger children simple things like how many more red shirts do you have than blue shirts or what color toy cars do you have the most of. For older children budgeting and shopping are always popular. Cooking is fun for any age. Younger children can work on measuring and older children can work on adding and dividing fractions to double the recipe or make half a recipe. Sale papers are excellent. Older children can use them to calculate the price in 'percent off sales'. Younger children can work on counting. They can also use the sales papers to determine the cost of several items or how much money they would have left if they bought a specific item.

For your child, first make sure homework is attempted. Children must learn to ask for help when needed. However, I know one of my own children always just tried to get out of the work with either delaying tactics or asking for help before they even tried the problem. That isn't helpful to anyone. Try to use humor when things are difficult. Be sure you are joking with the child in a way that can't be confused with laughing at the child. One of the most important learning activities is correcting mistakes. When your child gets a problem wrong, help him or her correct it following these steps. First, determine if it is a simple error and the child can make corrections alone. If so, simply let the child know they need to make corrections. Second, if the child doesn't understand, reteach/explain how to do the problem. This is one of the reasons a child should attempt a problem. It helps you to determine why they are getting the problem wrong. Often, I use a similar problem with different numbers and work through it with a student. Then the student does the original problem independently. It is important to provide needed support, but at some point, the student needs to work independently, or they won't learn the skill or concept. It is also important for children to know that correcting mistakes is how you learn. Getting problems right the first time is great, but that indicates they already knew how to do the problem, not that they learned how. Correcting mistakes indicates they are learning how to do a problem that they did not know how to solve. If your child is frustrated when asked to make corrections, be sure to explain this concept.

For the teacher, ask what she is doing for your child. Ask what strategies she has used and what she suggests for you to work on at home. Avoid special education classes. Sometimes, they are good, but most times students in special education classes learn even less. Everyone just expects less of them when what most students need is a different type of instruction. If your child is already in special education insist that there is a plan that helps the child to catch up. Insist that a child is taught in a different way than in the general education classroom. Bring up the fact that struggling learners need explicit instruction, not discovery model instruction. If the school refuses, ask to see data showing that your child is making progress towards closing the gap. If the data doesn't show the gap closing, point out that means the current program isn't working and you want it to change. Every child that is not mentally impaired can learn elementary level math if taught the way that is best for the child. Insist the plan for your child reflect that. For more information on Special Education, see my book on the problems with special education, to be published in 2018.

Another thing is that students need skills. The big thing about the interpretation of Common Core is the application of skills. Amazingly, it seems only a few educators realize that you can't apply skills you don't have. Teach your child the skills. They are the easiest to teach at the elementary level. Remember, that most people can do everyday math and your child will be able to as well with a little help. The students that get all the Common Core Math will be the ones to go into math fields. If your child is interested in one of those areas, then be sure to get the best help.

This brings me back to Kumon which I love. The founder is Japanese. When his son was young he started making little tasks for him to do every day so that he would be able to pass entrance exams as a teen. It was so successful, that other parents asked for help for their children and soon it was a booming business. I know when I take my child to Kumon, there are way more Asians there than any other ethnic group (we do not live in a predominantly Asian neighborhood). That says it all to me. Kumon teaches skills and the Asian kids are the ones that are at the top of the math classes at my daughter's high school. This says to me, Common Core or not, the skills are still important. Once a child masters the basic skills, it will be so much easier to apply those skills and to solve problems requiring higher level thinking. If your child is struggling with multiplication facts, there is no way they are going to easily solve or even understand complex problems involving fractions or ratios. If memory serves me correctly, Kumon teaches multiplication and division facts through the 20s, not the 10s which is the current standard in elementary school.

Finally, remember that when the math pendulum swings back the other way, to mastering skills along with applying concepts, your child will be ready to go because your child will know the skills and strategies needed to solve complex problems. You can't solve problems if you don't have skills. When your child gets to high school, they will be prepared for high school math if they have mastered math skills and concepts, even if they struggled with all the Common Core curriculum. To help your child survive the current fade in Common Core Math Instruction, focus on developing math vocabulary, mastering basic facts and learning skills and concepts.

Synopsis

Problems

1. What we think of as Common Core is really the Common Core Package which includes the Common Core Learning Standards, Educational Reform, Curriculum/Textbooks and High Stakes Testing.
2. Common Core Math Standards were created by math gurus who do not understand the average math student, much less the struggling math student. There are no provisions for filling in the gaps in math knowledge.
3. Math gurus' attempts to help students understand math really made math more confusing for the average student, the struggling student, and many parents.
4. One-size fits all math curriculum is being labeled as Common Core. This seems to have come into existence to sell math textbooks and professional development. It would be difficult to sell the new and expensive math programs with just minor changes that reflected the changes in Common Core Math. The Curriculum does not do enough to address low achievement levels, lack of prerequisite knowledge, and various learning styles.

Goals

1. We must keep in mind that the number one goal of school is education. Our children should be learning. Solutions to the problem should not impede the goal of learning.
2. Along those lines, the goal for the average student is to receive an education that gives them choices as adults. In other words, it prepares them for post-secondary education in college or a technical school without costly remedial math classes during college years.
3. Fostering self-esteem is essential to education. Students who think they can are much more successful than students who believe they cannot.
4. Finally, preserving our relationship with our children is imperative. We should not have to choose between having a good relationship with our children or helping our children be successful.

Solution

Essential Facts

- Math builds on prior knowledge
- Schools teach by grade level, not student achievement levels
- Students must work at their achievement levels to fill in the gaps in knowledge

Steps to Take:
1. Students are handicapped when they do not understand math vocabulary. It is imperative that our children know math terms, understand what they mean, and be able to use that knowledge to solve math problems.

2. Of course, we want our children to know the basic skills and concepts being taught at their current grade level. If not, they will continue to struggle as future curriculum assumes previous skills and concepts have been mastered.
3. Along this line, it is also imperative that students know all skills and concepts from previous grade levels. If they do not, we need to provide the help they need to master those skills and concepts. This task only becomes more difficult as the years pass, and more information needs to be learned. The good news is it is pretty easy to teach a 5th or 6th grader the math skills they did not learn in 1st or 2nd grade.
4. No matter how good the program, students will learn very little if they are not motivated and engaged. If our children do not possess intrinsic motivation, we need to supply extrinsic motivation until they develop intrinsic motivation. Motivation leads to engagement in the learning task and then actual learning.
5. Everyone needs to be accountable, parents, children and teachers. As parents, we need to be sure we are accountable, then hold children and their teachers accountable.

Teaching Math Facts

Teaching math facts is one of the easiest things to do in math. Your child should start with multiplication facts, then division. This is because multiplication and division are the basis of much higher math. Students can always count on their fingers for adding and subtracting, but students need to know multiplication and division to see the relationship in numbers. It requires some time and patience to find the teaching strategy by which your child learns best. I remember as a young teacher, I went to a workshop for professional development. Turns out the presentation was not what I was expecting. However, it was one of the most helpful workshops I have attended. The workshop was all about teaching math facts. The presentation was selling a product, but it was a good product and it is one of the resources here.

As the presentation started, the speaker pointed out that one of the biggest claims by parents and teachers is that certain kids have poor memory skills and are just not able to memorize facts. The speaker continued, saying that was ridiculous. Unless the child is getting lost every day and cannot find the class, they are able to learn and memorize facts. The problem is teaching strategies for those students who are struggling. Typically, in school and at home students are presented with 10 to 12 facts to learn. With that many facts, it is all too easy to confuse 8 x 6 and 8 x 7. The key is in presenting just a few facts at a time to students who are struggling.

Below are several strategies, including Rocket Math, which was the company presenting the workshop I went to. Choose a strategy. Try it with your child. If it is working, continue until your child knows the facts. If it isn't working, try a different strategy. As mentioned earlier, an extrinsic reward may be necessary. Especially if your child has struggled for years. Do whatever it takes. Make it fun. If your child has struggled for a long time, you might even start by telling the child that you know he or she is struggling, but that is because all kids learn differently, and no one was teaching the way he or she learns best. It is always good to add something about how hard it is for teachers because there are so many different ways to learn, they can't teach them all in school. You don't want to encourage a mindset of the child's problems being someone else's fault even if that is true. Avoid the blame game while preserving your child's self-esteem.

1. Computer Games – some kids learn facts using computer games. If your child is highly motivated by video games, this may work well. However, computer games do not work for all children.
2. Multiplication War Card Game – this game can be purchased on Amazon or from teacher stores. You can also make your own by taking the face cards out of a deck of cards. It is played just like the card game war, but instead of putting down 1 card each player puts down 2 cards and then multiplies. Highest number wins the round.
3. Multiplication Bingo – this game can be purchase through Amazon or teacher stores.
4. Flash Cards – you can purchase flash cards or make your own out of 3 x 5 cards. Facts should be on one side only to

make sorting easier. To start, pull out all the 0s. This includes 0 x ___ and ___ x 0). When the zeros are mastered, add the 1s to the deck. Again, that includes 1 x ___ and ___ x 1). Go through the deck of 0 and 1 facts. Put all the cards that are known (child gives answer quickly) in one pile and the ones that are missed in another pile. Review the ones that are missed. On the next day, just go through the missed cards. Add the ones that are now known to the pile from yesterday and review the missed cards. Continue until all the 0s and 1s are known. The next day go through the whole set of known cards. The following day add the 2s. (If your child struggles here, just add half of the 2s.) Continue this process until all the facts are known. The good news is by time you get to the 8s there are only 3 facts to learn – 8 x 8, 8 x9 and 9 x 8.

5. Rocket Math – This is a commercial program for teachers, but can easily be used at home. The cost was only $29 at the time of publication. Go to RocketMath.com to purchase. As a side point, this is a tried and true product. It has been on the market for years and is still being sold. This is because the product works. It is a paper and pencil product and it takes only a few minutes a day. Each day a student is asked to learn 4 facts, such as 3 x 4, 4 x 3, 2 x 6 and 6 x2. Your child should not move to the next sheet until they have mastered the current problems. Repeat a sheet until your child has learned the facts on the sheet. I find that most students progress quickly until Letter I, then they slow down. If necessary, practice the key facts

at other times during the day. If your child is resistant, add some small reward – a smiley face or a sticker usually works well. The chart in the Rocket Math program is motivational for other children.

6. Drawing – if your child does not understand the concept of multiplication or division, it should be taught along with the facts. You can do this by simply asking your child to draw a picture of 2 x 3. (Stick with low numbers as the concept is the same and it takes less time than drawing 8 x 12. You could also buy one of the inexpensive practice books for multiplication – make sure it is not just the facts. **I cannot emphasize this enough.** Your child must understand what multiplication and division are as they master the facts.

7. Reading – make sure that your child can read the problems. This is especially true for division. Again, an inexpensive workbook is an excellent tool to help.

Filling in the Gaps

I have already talked about the need for your child to learn prerequisite knowledge they have missed. This section is just a review for parents who want help to do this themselves. You can generally go to your public library and see the text books being used by the various grade levels in your district. From these books you can pull out the skills and concepts your child needs to know. There are also thick workbooks covering everything a 1st grader (other grades as well) needs to know. The problem with both of these methods is that they are time consuming, unless your child is in 1st grade. Time is of the essence as your child works to do grade level work and fill in missing work.

If you would like a quicker solution, I recommend *Drops in the Bucket* by Frog Publications. For less than $20 you can get a blackline master (permission granted to make copies for personal use) of all the math needed for a particular grade level in 60 pages. I usually start kids on Level B. Most things in level A are easy to do yourself, if you have a younger child. If you go to the Frog Publication website, you can see an example of a page to know what type of problems are being completed. Again, just like Rocket Math, this is a time-tested product. In addition, this is a true intervention as it is set up so that a student can complete a year's worth of material in less than a year. All the fluff is gone because your child does not have time for fluff if they are trying to catch up. All for less than the cost of 1 hour of tutoring. One note, with the changes in Common Core Standards, some standards are not taught at the same grade level as the Standards. That is ok. Your child is not working at his or her current grade level and they will need to know all the math presented in the program.

Here is the procedure I used when I used *Drops in the Bucket* in my classroom.

1. Use level for where you think your child is functioning, not current grade level.
2. If unsure of starting place, copy Lesson 10 from the book you choose. Have your child complete the page with no help. If your child gets 100% repeat with Lesson 20. Keep going in this manner till you find the right starting place. At whatever point your child doesn't get 100, go back. For Lesson 10, start on Lesson 1. For Lesson 20 start on Lesson 11, etc. If your child gets less than 70% on Lesson 10, try an easier book.

3. Once starting level is determined, copy each day's lesson rather than have your child work out of the book.
4. Have your child complete the lesson. Check it immediately. Have child make corrections immediately. If child missed more than 2 problems, repeat same page the next day.
5. If your child continues to miss the same problem, find supplemental material for extra practice. (Common Core Sheets is a good place to start looking for printables.)
6. If your child gets 100% on a **new** page for 2 days in a row, skip 5 pages.

Note: First attempt each day should be independent work unless your child needs help to read something.

Grading: Each page is numbered 1 to 10. Some numbers consist of just one problem. Others have multiple problems. If you child misses any problems, the whole number is wrong. It isn't about the score, it is about your child getting the practice needed to master the concept.

Motivation: I always used some small reward system. A star or a sticker for 100% on a page, even if it is the 3rd attempt. Informal assessment (Lessons 10, 20, 30 etc) should get a slightly larger reward, such as extra time on a math game website or extra playtime. Finishing the book calls for a celebration, such as a certificate or special treat. Some parents set up elaborate rewards systems where students earn tickets that can be traded in for treats, toys, movies, etc. It is really up to you and your family values.

Internet Sites

CCSS Math Vocabulary: http://www.ncesd.org/Page/983
This site has math vocabulary flashcards for students in kindergarten through high school. Words are arranged by grade level with definitions and colorful pictures.

Common Core Sheets: http://www.commoncoresheets.com
Type: Paper Pencil
Cost: Free
This site has work sheets for grades K to 6. It is organized by topic or grade level. There are 10 sheets for each skill. If a sheet is too long, have your child do half. If your child still hasn't mastered the concept, either have him repeat them (it is math, most kids won't memorize the answers and know they are repeating the sheet). I usually change the page number by writing 1A, 2A, 3A or 11, 12, 13 the second time a student goes through a set of practice sheets. This will depend on your child. I have found that most children tend to ignore the page number and it doesn't really matter.

IXL: https://www.ixl.com
Type: Online
Cost: $79 a year/$10 a month
This site has practice through high school. It is an excellent site. It is organized by grade level. It scores your child's work and gives immediate feedback. It also analyzes your child's work and gives information on their progress. In addition to math, it also has language arts, social studies and science. For parents, you can find the Common Core Standards by grade level, even if you have not paid for a membership.

Kahn Academy:
https://www.khanacademy.org/math/arithmetic
Type: Online
Cost: Free
This site has practice through high school. It is organized by topic. It has many videos and is great for students who need visual and/or oral instruction. It also covers many other subjects in addition to math. Students are able to work at their own pace and feel successful using this program.

Math Playground:
http://www.mathplayground.com/flashcards_timed.html
Type: Online
Cost: Free
This site has games children can play to help learn math facts. In addition, there are games for word problems, logic and game linked to Common Core standards. It includes a link for parents to look at the Common Core standards.

Scholastic, Inc:
http://www.scholastic.com/parents/activities-and-printables/printables/ages-8-10
Type: Online Printables
Cost: Free
Scholastic has various worksheets parents can print for their children. A pulldown menu asks for an age range and type of activities you are looking for.

Homemade Math Games:
http://www.learn-with-math-games.com/homemade-math-games.html
Type: Online Directions
Cost: Free
Various math games that are easy to make at home.

Cool Math Games: http://www.coolmath-games.com/
Type: Online
Cost: Free
This site covers many math skills. It is a favorite with my students and I often use it as a reward.

Fun Brain: http://www.funbrain.com/
Type: Online
Cost: Free
This site covers a range of academic topics. It was recommended to me by my granddaughter, so children enjoy the site. This is another site that could be used as a reward, however as a parent you would need to monitor which games were chosen rather than allow your child do whatever they want.

Common Core State Standards Initiative:
corestandards.org
Type: Online
Cost: Free
Lists Common Core Standards, has a comparison between Common Core and old standards.

National Council of Teachers of Math (NCTM) - Click on Link for Math Standards: nctm.org
Type: Online PDF of Standards
Cost: Free

Educational Leadership: *How Emotions Affect Learning*
http://www.ascd.org/publications/educational-leadership/oct94/vol52/num02/How-Emotions-Affect-Learning.aspx

Great Kids!: *The Role of Emotions in Learning*
http://www.greatschools.org/gk/articles/the-role-of-emotions-in-learning/

Science Daily: *How Emotions Influence Learning & Memory Processes in the Brain*
https://www.sciencedaily.com/releases/2015/08/150806091434.htm

There are many other sites on the internet. It is simple to do a search for 'online math games or practice.'

Vocabulary Lists

(This is not a complete list – check your child's math program for additional words)

Kindergarten Math Vocabulary

heavier	shorter	near
lighter	longer	minute
shape	taller	day
square	above	clock
circle	below	week
triangle	in front of	dime
rectangle	behind	nickel
cube	beside	penny
sphere	inside	pattern
more than	outside	alike
less than	number	match
equal to	decompose	attribute
first	equation	bigger
plus	object	littler
minus	sort	category
take away	whole	classify
corner	zero	quantity

First Grade Math Vocabulary

tally mark	even	cone
add	odd	cylinder
sum	more than	hexagon
addend	fewer than	pentagon
addition	less than	rectangular prism
subtract	greater than	line
difference	place value (100s)	order
coins	attribute	side
estimate	fraction	weight
alike	category	height
timeline	half of	length
number line	most	a.m.
compose	quantity of	p.m.
decompose	row	analog

Second Grade Math Vocabulary

symbols: < > = ≠ + - x array regroup multiply product difference sum numerator denominator place value (1000s) 1/2 1/4 1/3 fraction halves word form equal to estimate expression penny nickel dime quarter dollar key	clockwise data digit dozen fact family maximum minimum mode operations associative property-addition commutative property category customary system decompose compose key minute hour inch foot yard	2-D 3-D angle obtuse angle acute angle right angle plane prism sphere polygon quadrilateral face edge vertex-vertices perimeter congruent symmetry parallel meter centimeter

Third Grade Math Vocabulary

whole number	associative	isosceles triangle
decimal point	property + x	equilateral triangle
place value (tenths)	commutative	scalene triangle
sequence	property	area model
numerator	distributive	intersection
denominator	property	flip-reflection
second	digit	area
equal to	compose	axis
not equal to	decompose	capacity
divide	expression	closed figure
multiply	evaluate	metric system
dividend	expanded form	net
divisor	pattern	probability
decimal	factor	range
bar graph	equivalent	remainder
algorithm	fractions	rotation-turn
mean	equal groups	translation-slide
median	end point	volume
mode	1/6 1/8	
	numerator	
	denominator	
	mile	
	kilometer	

Fourth Grade Math Vocabulary

sum	weight	point
difference	ounce	ray
product	pound	line
quotient	ton	diagonal line
factor	gram	parallel line
factor pair	kilogram	perpendicular line
mixed fraction	pictograph	plane
proper fraction	coordinate system	protractor
improper fraction	ordered pair	degree
benchmark fractions	function table	rotation
simplest form	identity property	parallelogram
GCF	attribute	trapezoid
LCM	properties	rhombus
ratio	equation	quadrilateral
mixed number	inverse operation	rectangular prism
like denominator		acute angle
		obtuse angle
		right angle
		vertex

Fifth Grade Math Vocabulary

simplify	percent %	Congruence
common	estimate	volume
denominator	circle graph	irregular polygon
GCF	pictograph	polygon
LCM	line graph	capacity
improper fraction	pie chart	circumference
divisibility	line plot	diameter
multiply	cup	pi
factor	pint	axis
algorithm	quart	base
inequality	gallon	coordinate grid
algebraic rule	fluid ounce	mass
divisible	liter	radius
equation	milliliter	vertex
expression	cubic centimeter	ordered pair
exponent	meter	x-axis
prime number	cubic meter	x coordinate
prime factoring	cubic inch	y-axis
square root	cubic foot	y coordinate
variable	cubic yard	attribute
powers of ten	interval	ordered pair
tenths	frequency	
hundredths	properties	
thousandth		

Math Games

Math Board Games for Purchase
Dino Math Facts
Fish Stix
Head Full of Numbers
Math Dice Junior
Money Bags Coin Values
Monkey Math
Mountain Raiders
Multiplication Bingo
Sea 10
Sequence Numbers
Sum Swamp Game
Zingo Tells Time

Math Card Games for Purchase
24
Math War
Snap-It Up

Electronic Math Games for Purchase
Math Minute Electronic Fact Game
Cash Register
Osmo Gaming System for iPad
Play Math for Nintendo DS

Games that Use Math
Battleship
Connect 4
Life
Mastermind
Monopoly
Pay Day
Phase 10
Rummikub
Skip Bo
Sorry
Trouble
Yahtzee

Math Games To Make

Flash Cards

Math War: Use a deck of cards. Remove face cards. Choose operation – add, subtract, multiply or divide. Deal all the cards. Each player lays down 2 cards. The player with the highest value after performing operation wins all played cards. Player who wins all the cards or has the most cards when time is up, wins game.

Matching: Make 2 sets of cards, one with pictures and one with numbers. This will work for many types of skills such as fractions, counting, decimals, vocabulary, etc. This can be played by one child or a competitive game where the child with the most cards wins.

Books: Various books are available with games to make. Evan-Moor publishes Math with Games (for Primary students). There are others, but it is best to go to a book store or teachers store to look at the games they have to offer. No need to purchase a book of games that does not address any of the topics your child needs to work on.

Recommended Reading:

- Parenting with Love and Logic by Foster Cline and Jim Fay
- Boundaries with Kids by Henry Cloud and John Townsend

About the Author

Carol Pirog has been teaching in the public school system for many years. With all the confusion when Common Core was introduced, Carol set about determining what was really helpful in the Common Core with other veteran teachers at the low-income school where she taught for years. Carol has lived in a number of different states and has observed the elementary educational system and found it lacking for many students. Doing personal research, Carol has found a system that reverses the failure many students experienced for years. Carol lives in the Midwest with her family.

Look for Carol's new book in 2018 on the special education system in public schools.

Made in the USA
Middletown, DE
09 November 2018